How We Became Human
On Human Origins
and Our Path to Enlightenment
N. K. Sachdev

How We Became Human

How We Became Human

Copyright ©2019 by N. K. Sachdev

All rights reserved. No part of this book may be reproduced or transmitted in any form or by any means without written permission from the author.

Published in the USA

DEDICATION

This book is dedicated to the champions of humanity; and, in particular, to all those who have helped build the foundations of modern human society. It is also dedicated to anyone seeking to truly understand life, how humans came to be, and how we can achieve our greatest potential.

CONTENTS

   Acknowledgments

1 Introduction   1

2 Our Origins   5

3 Our Genetic Lineage   19

4 Beauty of the Human Form   23

5 Becoming Human   29

6 Balancing Instinct with Intellect   41

7 Carving a Path   45

8 Humanity's Spirit   49

9 Humanity's Potential   53

   Epilogue   57

ACKNOWLEDGMENTS

True strength can be measured by how we treat others, the planet, and ourselves.
—*Author Unknown*

The inspiration for this book is the quest to understand the meaning of life that is inherent within all living things; and, in particular, to understand how humanity has emerged. As humans, we seek to understand where we came from, how we became human, what is our role in this world, and what the future holds.

I spent many years trying to understand the human experience and thinking about and analyzing how humans came into existence. In the process, I also studied various cultures, belief systems and religions around the world. At the core of my findings was the commonality of life experiences and challenges of varied humans around the world despite their culture, belief system or religious background. The life experience of humans did not make sense in the absence of an evolutionary framework for our origins. This quest to understand nature, life, existence and humanity led me to seek answers to the penultimate question of where did we come from and how did we come to be? This question logically leads to another fundamental question, which is what is our role in life and what does the future hold? These are the questions that I attempt to address in this book.

I benefited greatly from knowing and studying various cultures and belief systems around the world, including my own experiences of being exposed to both Eastern and Western though and society. I also benefited from a keen passion for science, math, physics, and medicine.

To many, it is a true mystery how humanity has found its way on this planet. However, there are clues that hint at our origins and the path we have taken.

The story of humanity and its evolution is truly unprecedented. The findings presented in this book reflect the results of my in-depth inquiry into the human experience, including our evolution and our connection with nature. My sincere hope is that these findings help add to the human story in some way and help anyone seeking to make sense of, or to enhance, their human experience.

The intent of this book is to share the findings of my inquiry and to enlighten those who may come to see a similar path for humanity. In presenting the findings of this inquiry, I wish to acknowledge all those who prompted the inquiry. I would also like to acknowledge all individuals, both known and unknown, who, through their work in the fields of science, technology, engineering, medicine, law and a host of other disciplines, have helped to advance humanity's standing in the world. A debt of gratitude is owed, in particular, to those who have helped to advance the story of life as well as those who, through their ideas and actions, have helped improve the welfare of their fellow human as well as, in particular, those who have helped the welfare of humanity as a whole. Finally, immense respect and gratitude is owed to our ancestors who helped carve our path.

I also wish to acknowledge all noted references for this book. Any omissions in acknowledgement of authors and sources should not be taken as errors of commission; rather they should be considered errors of omission.

N. K. Sachdev
sohin9@gmail.com

1 Introduction

We are very, very small, but we are profoundly capable of very, very big things.
—*Stephen Hawking*

Humanity's origin is the subject of intense debate. The need to understand our origin is as old as time itself. Deciphering how we were formed is important not only for the purpose of gaining knowledge, it is crucial for understanding humanity's role in this world and for securing our future direction.

Since Charles Darwin published his theory of evolution in his book *"On the Origin of Species"* in 1859, humans have been searching for the missing link that allowed humanity to emerge. We don't look or act like chimpanzees or bonobos and yet we share nearly 99 of DNA with these ancestors. Archaeologists have found hominin (species closer to humans than chimpanzees) remains dating as far back as 4 million years and our pre-hominin ancestors are thought to have lived about 6 to 7 million years ago. However, we have not yet found the missing link. Although the possibility exists that there is a "missing link" yet to be found, there is clear evidence that a number of factors have shaped humanity. Understanding the factors that have contributed to our evolution does not substitute for a missing link. Nevertheless, knowing the various factors that have shaped humanity sheds extensive light on our origins and thus our future.

Our planet has undergone vast changes over the 4.54 billion years of existence; and, based on currently available evidence, modern humans—*Homo sapiens*—are thought to have been on this earth for about 200,000 years. Where did we come from? How did we come into existence? What has shaped our world, and our bodies and brains? In our quest for survival, we humans have markedly shaped our world. However, what has shaped us? What has played a key role in shaping our brains and bodies? How has this been accomplished?

Life and its survival on our planet is impacted by a number of variables, e.g., available raw materials (Goldilocks conditions), climate, geography, food sources, resourcefulness, adaptation, competition, predation, communication, collaboration, group strength, environment and genetics. Given a set of conditions, those life forms that are the fittest, and exhibit sufficient reproductive success survive and become ancestors. So, how have we remained fit and adaptive? What has contributed to our remarkable reproductive success? This, despite the fact that the human population came close to extinction a number of times. Some estimates indicated that the human population dropped to about 1,000 (one thousand) reproductive adults at one point, and there is speculation that the number may have declined to about

40 (forty) at the nadir. Today, the number of humans on the planet exceeds 7 (seven) billion and this count is expected to grow. Humans have found ways to survive despite the odds and have exhibited tremendous reproductive success.

This book aims to shed some light on how humanity has survived and thus our origins. It is not a scientific assessment. Instead, it offers a practical approach to understanding who are we and how we have likely evolved into our current selves. We are known as the the fifth ape or the naked, hairless ape. The other apes—orangutans, gorillas, chimpanzees, and bonobos—are our closest genetic relatives. How have we diverged from the path of these apes and emerged as the naked, hairless ape and look and operate like we do today? What are the forces at play that have created us, who are called *Homo sapiens* or "wise men"?

2 OUR ORIGINS

Chimps are unbelievably like us — in biological, non-verbal ways. They can be loving and compassionate and yet they have a dark side…98 per cent of our DNA is the same. The difference is that we have developed language — we can teach about things that aren't there, plan for the future, discuss, share ideas…
—Jane Goodall

As humans we naturally want to understand our origins—where we came from and how we came into existence. Life on earth and, in particular our human origins, has been the subject of intense debate since Charles Darwin proposed that *Homo sapiens* are the result of evolution from other life forms. According to Darwin, not all living beings co-existed on earth from the beginning, but with changing conditions, living beings that adapted to these changes survived. Darwin postulated that humans evolved over time, starting with life forms in the water, to the amphibians, and then to apes and then half man-half ape, continuing onto becoming *Homo sapiens*. He described humanity's origins as part of a larger tree of life. Today, indeed there is clear evidence of, and support for, the common descent of all living organisms among scientific researchers in multiple disciplines.

DNA studies indicate that the human family originated in Africa. Pre-hominins date back to about 6 to 7 million years ago and early hominins are thought to have existed some 4 millions years ago. Modern humans or *Homo sapiens* are thought to have emerged about 200,000 years ago, which represents only a small fraction of time of our planet's history. How we reached our present day form remains somewhat of a mystery.

A host of variables likely interplayed in shaping humanity. Some of the key variables among these include: walking upright and bipedalism, controlling fire, bigger brains, tool making, and a shift in diet.

Walking Upright and Bipedalism

Among the distinguishing characteristics that separate humans from other mammals is walking upright or erect bipedally. Walking upright is thought to be among the first key steps toward humanity. Indeed there are other bipedal animals alive today including birds. However, neither birds nor such animals walk upright on land like humans. Walking upright is a form of bipedalism, a feature that humans are thought to have developed even before they developed bigger brains.

Our bipedal origins are documented in the fossil record. The Ardi fossil, discovered in Aramis, Ethiopia in 1994, was recovered over a number of years after its teeth and small bones were initially found in 1992. The Ardi remains date back 4.4 million years. Ardi is classified as a complete early hominid specimen comprising most of the skull, teeth, pelvis, hands and feet. Ardi, thought to be an early human-like anthropoid, likely lived on trees as well as on land. Though Ardi is thought to be bipedal, she had opposable big toes and thumbs in order to be able to climb trees. Ardi's bipedal movement was likely impeded, but it is thought that it allowed Ardi to bear additional offspring. Although it is not known if Ardi's species, *Ardipithecus ramidus*, led to the development of *Homo sapiens*, Ardi represents a transitional form between tree and land dwelling. Prior to the Ardi specimen, a collection of fossilized bones called Lucy was discovered in 1974 in Ethiopia (approximately 74 kilometers from the Ardi specimen). Lucy, who lived about 1.2 million years after Ardi, i.e., approximately 3.2 million years ago, is thought to be the first ape that walked upright. Another find, the Taung child (who lived approximately 0.9 million years after Lucy), which is the fossilized skull of a child found in 1924 in Taung, South Africa, dates back to 2.3 million years ago and is considered to be of the genus Australopithecus. After extensive debate, Taung child, which was a transitional form between apes and humans, was declared a hominid (early human) as it was human-like in its posture, dental structure, and bipedal walk. *Homo erectus*, discovered in 1991 in the country of Georgia, dates back to 1.8 million years ago. *Homo erectus* is considered to be the first human to have walked fully upright and possessed human-like body proportions. Thus, based on the existing fossil record, there is clear evidence that bipedalism evolved early on in the development leading to us—*Homo sapiens*.

There are clear advantages for humans due to the hands being free while the legs are used for walking, running, etc. Some of the advantages include ease of carrying things, digging, managing and controlling fire, cooking, using tools, playing an instrument, drawing, writing, biking, driving, etc. Without bipedalism and the resultant freeing up of the upper body and the arms and hands, humans could not exist as we do today nor could we engage in the multitude of professions (e.g., home builders, surgeons, dentists, firemen, teachers, writers, scientists to name a few) and activities in which we participate (e.g., cooking, biking, walking, driving). Bipedalism has offered humans a clear advantage as compared with other animals. This includes the exceptionally advanced ability to grab, hold, carry, pick up, throw, touch, and see from a higher vantage point. Indeed, the list of benefits that bipedalism has offered to humans is substantial.

The question is how did we become bipedal? There are many theories, including the fact that walking upright uses fewer calories than walking on all fours. Some of the existing theories are discussed below:

As reported in the Smithsonian magazine:

> "... in the 1920s when anatomist Raymond Dart discovered the skull known as the Taung Child in South Africa. Taung Child had a small brain, and many researchers thought the approximately three-million-year-old Taung was merely an ape. But one feature stood out as being human-like. The foramen magnum, the hole through which the spinal cord leaves the head, was positioned further forward under the skull than an ape's, indicating that Taung held its head erect and therefore likely walked upright. In the 1930s and 1940s, further fossil discoveries of bipedal apes that predated Neanderthals and *H. erectus* (collectively called australopithecines) helped convince anthropologists that walking upright came before big brains in the evolution

of humans. This was demonstrated most impressively in 1974 with the finding of Lucy, a nearly complete australopithecine skeleton. Although Lucy was small, she had the anatomy of a biped, including a broad pelvis and thigh bones that angled in toward the knees, which brings the feet in line with the body's center of gravity and creates stability while walking.

In more recent decades, anthropologists have determined that bipedalism has very ancient roots. In 2001, a group of French paleoanthropologists unearthed the seven-million-year-olds *Sahelanthropus tchadensis* in Chad. Known only from a skull and teeth, *Sahelanthropus'* status as an upright walker is based solely on the placement of its foramen magnum, and many anthropologists remain skeptical about the species' form of locomotion. In 2000, paleoanthropologists working in Kenya found the teeth and two thigh bones of the six-million-year-old *Orrorin tugenensis*. The shape of the thigh bones confirms *Orrorin* was bipedal. The earliest hominid with the most extensive evidence for bipedalism is the one called Ardi.

Although the earliest hominids were capable of upright walking, they probably didn't get around exactly as we do today. They retained primitive features—such as long, curved fingers and toes as well as longer arms and shorter legs—that indicate they spent time in trees. It's not until the emergence of *H. erectus* 1.89 million years ago that hominids grew tall, evolved long legs and became completely terrestrial creatures.

While the timeline of the evolution of upright walking is well understood, why hominids took their first bipedal steps is not. In 1871, Charles Darwin offered an explanation in his book *"The Descent of Man, and Selection in Relation to Sex."* He noted that hominids needed to walk on two legs to free up their hands. He wrote that '...the hands and arms could hardly have become perfect enough to have manufactured weapons, or to have hurled stones and spears with a true aim, as long as they were habitually used for locomotion.' One problem with this idea is that the earliest stone tools don't show up in the archaeological record until roughly 2.5 million years ago, about 4.5 million years after bipedalism's origin.

But after the unveiling of Ardi in 2009, anthropologist C. Owen Lovejoy of Kent State University revived Darwin's explanation by tying bipedalism to the origin of monogamy. I wrote about Lovejoy's hypothesis for *EARTH* magazine in 2010. Lovejoy begins by noting that Ardi's discoverers say the species lived in a forest. As climatic changes made African forests more seasonal and variable environments, it would have become harder and more time-consuming for individuals to find food. This would have been especially difficult for females raising offspring. At this point, Lovejoy suggests, a mutually beneficial arrangement evolved: Males gathered food for females and their young and in return females mated exclusively with their providers. To be successful providers, males needed their arms and hands free to carry food, and thus bipedalism evolved. This scenario, as with the various hypotheses on the origins of bipedalism, is really hard to test. But earlier this year, researchers offered some support when they found that chimpanzees tend to walk bipedally when carrying rare or valuable foods.

Another theory suggested that hominids evolved to walk upright in response to climate change. As forests shrank, our ancestors found themselves descending from the trees to walk across stretches of grassland that separated forest patches. The most energetically efficient way to walk on the ground was bipedally, Rodman and McHenry argued. (Full disclosure: Rodman was my graduate school advisor)..."

"Many experts believe that the key to the creation of the broken, mixed environment of trees, shrubs and savannah grasslands in Africa in which *afarensis* evolved, is the monsoon. Many millions of years ago, India collided with the continent of Asia, buckling the Earth to create the huge mountain range that we call the Himalayas.

This geological event created a monsoon that released vast quantities of rain, drying out the air. This same air flows across East Africa, and causes rainfall to drop sharply, drying ancestors found themselves descending from the trees to walk across stretches of grassland that separated forest patches. It is no coincidence that the monsoon intensified 6-8 million years ago--the time at which the common ancestor lived."

So why did bipedal apes emerge in this environment?

"Paleoanthropologists have advanced many theories over the years as to why quadrupedal (four-legged) apes began walking upright. One of the most recent, and convincing, theories has been suggested by Dr. Patricia Kramer of the University of Washington, who believes that bipedalism was the most energy-efficient way of moving around the broken landscape that appeared in Africa between 6 and 8 million years ago. Energy is critical a commodity to waste if you are a primate living a marginal existence in the African savannah. Its conservation becomes critical. Dr. Kramer found that the "short-legged morphology of early bipeds was the most energy-efficient body shape for covering relatively long daily distances on the ground. Once bipedalism took hold, there was no looking back."

"Related research findings indicate that bipedalism also reduces exposure to the sun, keeping the body cooler. This allowed our ancestors to be active mid-day when it was too hot for both predators and competitors to be out and about, giving us a huge advantage.

Numerous other explanations for bipedalism have been outright rejected, such as the idea that our ancestors needed to stand up to see over tall grass or to minimize the amount of the body exposed to the sun in a treeless savannah. Both ideas were debunked by the fact that the first hominids lived in at least partially wooded habitats."

Becoming bipedal clearly made it easier for our ancestors to tame fire, cook food, and more easily digest food from the esophagus into the stomach. It also offered a more energy efficient method of movement and increased energy available for the growing human brain. While there are drawbacks of bipedalism, e.g., back pain, knee issues, the benefits clearly outweigh the costs.

Controlling Fire

Another aspect that separates humans from other animals is the ability to manage and control fire. Humans are thought to have discovered fire about 1.6 million years ago. However, being able to control fire is thought to have occurred later—some 600,000 to 400,000 years ago. Widespread use of fire by humans is, however, estimated to have begun about 120,000 years ago.

Being able to tame and control fire gave our ancestors an extraordinary advantage. Fire allowed them to be safe from wild animals, particularly at night, and it served as a source of warmth and light. In addition, it offered them the ability to cook food. Cooking would have served to not only expand the food sources available to our ancestors, the process of cooking concentrated the energy imparted by the food, and it served to reduce pathogens in the food supply. Fire served a numbers of other functions for early humans. In conjunction, fire served to establish strong social groups among early humans who sat together by the fire to eat, to keep warm, and to shares stories and experiences with each other. Fire served as a powerful energy source and a social group connector for humans to expand their reach across the world and to better control their environment. In the process, it helped mold the bodies and brains of our ancestors.

The following excerpt reported by Tim Redford in "The Untold Story of Evolution" reflects scientific thinking regarding fire and cooking and its impact on our bodies:

"Did humans discover the use of fire millions of years ago, long before the colonisation of Europe? Cooking would make plants both more nourishing and easier to digest; it would dispose of infections and pathogens in meat, and it would deliver greater supplies of energy per mouthful. Teeth, jaws and digestive tracts could shrink, and so brains could get bigger. Did humans grow bigger brains because the extra neural circuitry was needed to make sense of the demands of social and co-operative life?

According to K Kris Hirst:

The discovery of fire, or, more precisely, the innovation of the controlled use of fire was, of necessity, one of the earliest of human discoveries. Fire's purposes are multiple, such as to add light and heat to the nights, to cook plants and animals, to clear forests for planting, to heat-treat stone for making stone tools, to keep predator animals away, to burn clay for ceramic objects. Undeniably, there are social purposes as well: as gathering places, as beacons for those away from camp, and as spaces for special activities.

The human control of fire likely required a cognitive ability to conceptualize the idea of fire, which itself has been recognized in chimpanzees; great apes have been known to prefer cooked foods, so the very great age of the earliest human fire experimentation should not come as a terrific surprise.

Archaeologist JAJ Gowlett offers this general outline for the development of fire use: opportunistic use of fire from natural occurrences (lightning strikes, meteor impacts, etc); limited conservation of fires lit by natural occurrences, using animal dung or other slow-burning substances to maintain fires in wet or cold seasons; and kindled fire. For the development of fire's use, Gowlett suggests: using natural fire events as opportunities to forage for resources in landscapes; creating social/domestic hearth fires; and finally, using fires as tools to make pottery and heat-treat stone tool.

Fire allowed early humans to survive and it helped facilitate the development of bigger brains. Given the role that fire played in our history, it is no surprise that many religions around the world, particularly older religions, still use fire ceremoniously and as a purifier?

Bigger Brains

Humans possess one of the largest brains on the planet; although, there are other animals with larger brains. In fact, the sperm whale is thought to have the largest brain among creatures alive today. Elephants are also endowed with a fairly large brain. Crows, ravens and magpies have fairly large brains relative to their body size and are considered as intelligent as primates. However, when compared to all other vertebrates, humans possess the largest brain relative to their body size. The human brain weighs about 3.3 pounds (1.5 kilograms) and the brain-to-body mass ratio is 1:50. Scientific studies show a progressive increase in brain size relative to body size as pre-hominids evolved into hominids and ultimately became *Homo sapiens*. Human brain size has been trending upwards since 2 million years ago, with a 3 factor increase over the period.[1] The human brain size increased most rapidly between 800,000 to 200,000 years ago. This was also a time of dramatic climate change. As the climate change made the environment more unpredictable, larger brains are thought to have helped our ancestors survive.[2]

Brain size alone, however, is not the best predictor of advanced cognition. Structural development of the brain is important for increased cognition. Studies on the structural components of the human brain also show evolutionary advances as compared to other vertebrates. Jelly fish had the first nerves and the first nerve led to the first spinal cord, which formed the first vertebrates. The nerves of all animals are the same. The difference is the level of structure and organization. There are three main regions of an advanced brain—the forebrain or the cerebrum, the midbrain or the cerebellum, and the hindbrain the major part of which comprise the hippocampus and the amygdala. Although brain functions in humans are highly complex, the hindbrain is primarily focused on supporting basic survival, e.g., respiration, blood flow, etc., while the midbrain focuses on motor control. The cerebrum, which is responsible for the higher cognitive functions, is the newest among the three regions in terms of evolutionary development. It first developed some 200 million years ago. This part of the brain, called the forebrain, exhibits the greatest amount of recent evolutionary change. The cerebrum is a large part of the brain that includes the cerebral cortex, including the gray matter and the left and right hemispheres. The cerebral cortex is the seat of higher brain functions—in particular, it is responsible for integration of information and formulating associations.

Compared to other mammals, the difference that stands out the most in humans is the size of the cerebral cortex, which is highly expanded and comprises folded cortices.[3] Other animals with highly expanded and folded cortices include dolphins and elephants. Recent studies show that the brain has ancient origins

—the common ancestor of annelid worms, insects and vertebrates possessed primitive brain structures. Bigger brains are needed in high level animals as the amount of information needed to be stored, analyzed and processed exceeds existing capacity. With the passage of time, the progressive rise in intelligence of our ancestors reached the point where, for the first time, the human line possessed the required brain structures. The boost in intelligence and cognition gave our ancestors an evolutionary edge.

Tool Making

Our ancestors excelled as tool makers. Tool making, however, is not something that sets humans apart from animals. Even octopi are know to use tools and have ape-like minds. However, the means and the complexity of tool making sets humans apart from other creatures. The forethought and rigor with which our ancestors built and utilized tools (and modern humans continue to do today) is unparalleled in the animal kingdom.

Based on archeological findings, stone tool making by humans dates back about 2.6 million years ago. The first such tools was the Oldwan—a stone tool made of sharp flakes by striking the stone against quartz or another stone to form a sharp edge. Over time, tool making evolved such that rough hand axes and cleavers were in use 1.8 million years ago, which is about the time *Homo erectus* emerged.[4]

Although humans are not the only living beings that make tools, humans have developed comparatively highly advanced tool making capabilities. Studies show that many animals use various tools, in particular for accessing food. For examples, crows use various techniques to access food and water. As noted in one of Aesop's tales, crows can quench their thirst by throwing stones into a pitcher to displace water and thus raise the level of water. Monkeys are known to use twigs to pull out termites and ants from a hole. Chimps can make their own spears for hunting. There are many ways other animals, including apes, elephants, dolphins and octopuses build and use tools. However, the complexity and array of tools made and used by humans sets humanity apart from all other animals.

Humanity has benefited from a host of capabilities in concert with toolmaking. These include the ability to copy or mimic actions and behaviors, e.g., observing another human use an axe and copying it. Moreover, humans have excelled in learning, practicing and, in particular, teaching other humans. In particular, humans have excelled in forming complex social relationships and communication channels using various tools and techniques.

Dietary Shift

As mentioned, cooked food dramatically expanded our ancestors' food sources and the level of energy derived from food. It also also allowed humans the ability to consume meat. In concert, the use of fire reduced bacterial and fungal infections through the use of heat to disinfect the food and water supply. The impact on humanity through the consumption of cooked food (in conjunction with raw food) is substantial. As noted by scientists, consuming primarily cooked food likely allowed our ancestors' digestive tract and stomach to contract, shaped our digestive system, and offered energy to support the advanced brain. Cooked food also allowed our ancestors to better store food for future use and thus freed humans from worrying about the next meal to some extent. This freedom helped pave the way for humans to focus on issues other than daily survival and helped build the foundations of modern humanity.

3 OUR GENETIC LINEAGE

DNA is like a computer program but far, far more advanced than any computer program ever created.
—*Bill Gates*

Modern humans or *Homo sapiens* are estimated to have lived in Africa nearly 200,000 years ago. This finding is supported by genetic studies. Genetic mapping has revealed that all humans alive today are genetically linked to a common ancestor for males and females, respectively.

The common male and female ancestors of present day humans have been identified. Based on genetic research, y-chromosomal Adam has been identified as the patrilineal most recent common ancestor (y-MRCA) and mitochondrial Eve has been identified as the matrilineal most recent common ancestor (mt-MRCA) of all currently living humans.[5] These genetic human ancestors are thought to have lived between 200,000 and 180,000 years ago. Although mitochondrial Eve and y-chromosomal Adam may not have lived at the exact same time, the genetic findings point to a common ancestral lineage for humans.[6]

The term y-MRCA reflects the fact that the y-chromosome of all currently living males is derived from and thus links back to a common male ancestor. Similarly, all living females genetically point to a common matrilineal (female lineage) ancestor. It should be noted that the terms "Adam" and "Eve" used by scientists in this context are not intended for any religious reasons, but to reflect the genetically inferred ancestral linkages for the modern human male and female that have been identified.

Furthermore, the studies do not indicate that "mitochondrial Eve" was the only woman alive at the time to which the human line has been traced. Indeed there would have been other females in existence. Scientists also state that "the definition of mitochondrial Eve is fixed, but the woman in prehistory who fits this definition can change." That is, not only can our knowledge of when and where mitochondrial Eve lived change due to new discoveries, but the actual mitochondrial Eve can change. "The mitochondrial Eve can change, when a mother-daughter line comes to an end by chance. On the basis of this definition of mitochondrial Eve, she had at least two daughters who both have unbroken female lineages that have survived to the present day. In every generation, mitochondrial lineages end when a woman with unique mtDNA dies with no daughters. When the mitochondrial lineages of daughters of mitochondrial Eve die out, then the title of 'Mitochondrial Eve' shifts forward from the remaining daughter through her matrilineal descendants, until the first descendant is reached who had two or more daughters who together have all living humans as their matrilineal descendants.

Once a lineage has died out it is irretrievably lost and this mechanism can thus only shift the title of 'Mitochondrial Eve' forward in time." 7

The "human tree" has been recreated using genetic mapping and it has been traced back down to a common base, but the base could change from what is currently known. Further mapping of living DNA lines could point the linkages back to an earlier woman." This happened to her male counterpart, "Y-chromosomal Adam," when older Y lines from Africa were discovered." Genetic studies also indicate that assuming that mitochondrial Eve lived at the same time as Y-chromosomal Adam (from whom all living men humans are descended) and perhaps even met and mated with him is unlikely. If this indeed is the case, it would be coincidental.

A brief timeline of human prehistory from the Middle Paleolithic era, i.e., when the first modern humans would have lived, is as follows:

Timeline of Human Prehistory: Middle Paleolithic Period

- 200,000 years ago: first appearance of *Homo sapiens* in Africa.
- 200,000–180,000 years ago: time of mitochondrial Eve and Y-chromosomal Adam.
- 195,000 years ago: oldest *Homo sapiens* fossil—from Omo, Ethiopia.
- 170,000 years ago: humans are wearing clothing by this date.
- 125,000 years ago: peak of the Eemian interglacial period.
- 120,000–90,000 years ago: Abbassia Pluvial in North Africa—the Sahara desert region is wet and fertile.
- 82,000 years ago: small perforated seashell beads from

Taforalt in Morocco are the earliest evidence of personal adornment found anywhere in the world.
- 75,000 years ago: Toba Volcano super-eruption.
- 70,000 years ago: earliest example of abstract art or symbolic art from Blombos Cave, South Africa—stones engraved with grid or cross-hatch patterns.
- 64,000 years ago: It has been speculated that the bow and arrow may have existed at this time.

Source: Wikipedia

Note: All dates are approximate and based on research in the fields of anthropology, archaeology, genetics, geology, and linguistics. They are all subject to revision based on new discoveries or analyses.

The genetic findings raise some key questions regarding how all humans alive today came from these genetic ancestors. How did the human genetic "tree" form along its path?

4 BEAUTY OF THE HUMAN FORM

Beauty lies in the eyes of the beholder.
—Plato

As said, we humans do not look like our genetic cousins. How did we come to be? Are we the mere result of bipedalism, use of fire, tool making, eating primarily cooked food, etc., or are there other precipitating factors that have given rise to humanity?

Returning to evolutionary theory, all living things small and big that exist are subject to natural selection. This is reflected in Charles Darwin's theory of evolution on survival of the fittest presented in his book *"On the Origin of Species"* published in 1859. Later, Charles Darwin presented his perspectives on sexual selection in the book titled *"The Descent of Man and Selection in Relation to Sex"* published in 1871. In this latter book, he concluded that the desire for beauty, i. e., pleasure itself, is the driving force behind the selection of the dances and songs of birds and that the same is true for humans. This second theory, which focuses on sexual selection and the concept of pleasure and beauty and its role in our world, is the subject of intense debate.

In trying to understand beauty in nature, the world of flowers offers some perspectives. Flowers vary in color, size, fragrance, texture, design, nectar, etc., in order to attract the right insect as the pollinator. There is a mutually beneficial relationship—the insect gets the nectar from the flower and it, in return, helps the flower to propagate itself. Although humans appreciate the beauty and the color of various flowers, this is what these plants must do to attract their intended audience—the pollinator. According to one evolutionary biologist:[8]

> People love flowers for their array of colors, textures, shapes and fragrances. But is pleasing the human eye the purpose of nature's floral design? Hardly. Survival is the plant's top priority, reminds Claude dePamphilis, a Penn State plant evolutionary biologist and principal investigator of the Floral Genome Project.
>
> "The beauty of the flower is a byproduct of what it takes for the plant to attract pollinators," said dePamphilis." The features that we appreciate are cues to pollinators that there are rewards to be found in the flower."
>
> Scent, color and size all attract a diversity of pollinators, which include thousands of species of bees, wasps, butterflies, moths and beetles, as well as

vertebrates such as birds and bats.

Flying insects, comprising the vast majority of pollinators, stop at the plant to eat nectar and pick up pollen, which they then distribute as they visit additional flowers. Noted dePamphilis,
"Pollinators are providing a very important service to the plant without which it couldn't reproduce."

To aid insects in finding the nectar — and thus, the pollen — many flowering plants have evolved to possess bright colors (hummingbirds and butterflies favor reds and yellows), as well as "nectar guides" that may only be visible in ultraviolet (UV) light — a wavelength of the light spectrum bees can see and people cannot. From a bee's-eye-view, the UV colors and patterns in a flower's petals dramatically announce the flower's stash of nectar and pollen.

The patterns on flowers that both humans and pollinators can see — such as the lines on petals called striations — serve as a sort of air traffic control system for bees, and help guide them into the "bull's-eye" of nectar and pollen at the flower's center, added dePamphilis. Thanks to this co-evolutionary trait that developed between the two species, the bee can efficiently visit many blossoms and pollinate a larger number of plants.

Some flowers, such as horse chestnut and sunflower, change colors within the ultraviolet spectrum throughout their lifespan. As dePamphilis explained, to pollinators these changes are visual signals of an abundance or lack of nectar, blaring "Visit me now!" or "Don't bother!"

To discover more about how these relationships evolved, dePamphilis and colleagues are using DNA-sequencing to dig back into the evolutionary history of flowers. "What we're really trying to do is infer the characteristics in a detailed, genetic way of what the earliest flowering plants were like," dePamphilis explained.

The fossil record dates the first flower somewhere between 125 and 140 million years ago. "How did those first flowers become the thousands of different varieties we now see in nature?" dePamphilis asked. "Gene mapping may shed some light on the beginning of the story."

While most flowers look and smell good to humans, some evolved strategies to attract their pollinators that are downright repellent to people, he noted. Plants belonging to the parasitic genus Rafflesia, native to Indonesia, produce what experts agree is the world's largest single flower, growing up to 3 feet wide. In addition to its sheer size, a Rafflesia flower announces its presence by its odor, a putrid stench of rotting meat. Though repulsive, the odor of "the stinking corpse lily," as the flowering Rafflesia arnoldii is known, proves irresistible to its main pollinators — carrion flies.

The array of fruits produced by various plants and trees also offers perspectives on beauty and how nature operates.

Many animals are attracted to the fruits, pods, berries, produced by particular plants and trees. The "fruit bearing" plants invite animals to eat the fruit and thereby help the plant spread the seeds contained in the fruit, pod, berries, etc. In the process, the plants and trees, like the flowers, have evolved to produce "fruits" of various colors, sizes, textures, aromas, etc., to attract the right animals (and humans). These fruits are colorful and beautiful and yet they are produced to attract the right creature to eat it and to help spread its seeds.

Looking at birds, the songs and the elaborate courtship displays of various birds reflect beauty. In the case of birds, although the relationship is not about a mutually beneficial relationship between two different species such as flowers and insects or plants and fruits, there is a focus on intra-species attraction and beauty in the context of sexual selection. Extensive research on birds and their activities has been conducted by Dr. Richard Prum, a renowned ornithologist. In his book, *"The Evolution of Beauty: How Darwin's Forgotten Theory of Mate Choice Shapes the Animal World —and Us"* (2017), Dr. Prum embraces Charles Darwin's "forgotten" theory of sexual selection. In accordance with this theory, sexual selection is a separate evolutionary force, driven by arbitrary aesthetic choices rather than by environmental imperatives, that drives natural selection. This second evolutionary force, sexual selection, according to Darwin, plays a critical role in shaping all species, including the human species. Thus, both natural selection and sexual selection interplay to shape all species.

Dr. Prum notes that Darwin's theory of sexual selection had two components: (i) male-male competition for access to females, and (ii) female selection of males based on preference for male behavioral and physical traits. The idea of male-male competition is not considered controversial, but female choice has often been dismissed, ignored, or presumed to be a variant of natural selection.

Dr. Prum refers to Darwin's description of females as having a "taste for the beautiful" meaning an "aesthetic faculty"and males as trying to "charm" their mates. Dr Prum notes that in a world were males operate on sexual coercion, female mate choice acts to counteract male coercion with female sexual autonomy. Dr. Prum supports his arguments with empirical evidence, mostly about birds. He argues that in humans, both the males and the females are selective and thus it is the selection process that produces (and shapes) the progeny. According to him, this selection process has shaped the human form.

Dr. Prum's perspectives on beauty and its evolution are controversial but provocative. Regardless of how the human form evolved, it is unequivocally accepted that it is dramatically different from its four-legged, hairy, quadruped cousins. The human form is truly elegant and beautiful. Perhaps, as Dr. Prum notes, all living things in nature have evolved to be beautiful to themselves. In other words, all creatures are attractive to whom they wish to attract for mating.

The forces that have created humanity are truly a wonder as there is no other animal quite like us—a naked, hairless bipedal primate with a big brain, born relatively helpless, in need of years of care; and yet, grows up to be a truly formidable force that wields tremendous influence and has learned to thrive across the planet. Indeed, in order for ancestral lineage to form, the fittest must survive and also reproduce. Humans have excelled in both respects. How this been accomplished is truly amazing.

5 BECOMING HUMAN

"The fully developed man changed from being savage and wild with the axe (Parasurama) to the spiritual man in the form of the Buddha."
—*Author Unknown*

Extensive studies indicate that great apes such as chimpanzees and bonobos are our genetic cousins. We share 98.8 percent of our DNA with them. So, how did we come to be and how did we "form" our features, our functions, our brains, and our bodies?. How did the human genetic tree form? We do not look anything like our closest cousins. We are hairless compared to our genetic cousins, we walk fully upright, we have bigger and more complex brains, and we eat a wide variety of foods, including both uncooked and cooked foods. (Even the poorest humans eat some cooked food). Humans also have a highly developed social culture, apply advanced communications methods (e.g., language), and utilize an ever growing array of tools and capabilities. Humans are part of the animal kingdom—more specifically, we are primates and we belong to the class named mammals and carry the higher designation of *Homo sapiens*. How did we come to be? Are we the mere product of natural selection and sexual selection?

The great apes benefited by moving from living on trees to living on the land (and simultaneously using the trees for safety). Similarly, our ancestors transitioned from tree living to living on the broken, wooded landscape. This adaptation was likely due to climate change—with the changing landscape (which would have meant fewer trees and limited fruit on trees) and the consequential need for supplementary food sources on the ground. Those ancestors who were able to adapt to living on the land benefited.

Bipedalism was crucial for our ancestors' survival in an environment of sustained climate change. It allowed them to survive while shifting to living on the land. Bipeds benefited from walking and running efficiently in this environment as they would have had to move fastidiously across wooded landscape as well as being able to climb rocks and cross jagged terrains, including climbing cliffs, to find safety in the caves.

As the author of this book, I consider that having hands and walking on two feet was a prodigious asset to our ancestors in such an environment. Let me explain. Living and nesting on land, unlike in the trees, our ancestors would have faced constant adversity—they would have had to fend off a host of new predators, e.g., lions, tigers, leopards, wolves and other wild animals in their midst, in order to survive. Also, they would have also had to deal with many poisonous snakes and a host of other creatures big and small. In this environment, our bipedal ancestors would have benefited from the efficient ability to move (and run) on the broken landscape to find hiding places in caves and other enclosed spaces, and to use their hands in a number of ways including for moving large rocks to block the entrances and openings of their "safe haven" places from wild animals. Else, these ancestors would have perished. Many of our ancestors probably did not make it alive on land unless they found safe hiding places in caves for extended periods of time. Cave dwellings would have also offered a natural shelter (a sanctuary) from the harsh climate. Most importantly, the bipeds, with their ability to walk upright would have benefited in being able to move efficiently, as well as being able to simultaneously carry their young and their foraged food (and eventually also their fire fuel in the form of dried twigs and branches) from the trees and from the ground with them into the caves and other sheltering places for extended periods of stay. Over time, the caves became their homes and their places of worship.

Furthermore, the bipeds would have been advantaged in being able to use their hands to create and ultimately control fire to ward off wild animals, to keep warm during long periods of cold weather, to have a source of light and safety, and to cook food (thereby increasing the nutritional density of their food and simultaneously reducing their exposure to food-borne viruses, molds, and bacteria). In the end, bipedalism would have also helped our ancestors to exploit multiple environments across the globe and expand their food sources, and thus paved the way for enhanced brain development. Bipedalism would have, in particular, paved the way for our ancestors to expand their horizons, e.g., making and using tools with greater ease, using fire to further expand food sources, getting rid of pests (by periodically burning the beds that they made using leaves and plant material to rid them of lice and other bugs), and exploiting resources in their surroundings to communicate with their peers about what they saw around them and perceived, and to warn them about the dangers they collectively faced—e.g., by making etchings on cave walls. Access to cooked food, particularly roots, tubers, and other plant material, would have also spawned further brain development, given the high energy offered by such foods.[9]

Research shows that brain development is experiential based. The greater the experiences that the brain undergoes and survives, the greater the development. Thus, having left the safety of the trees, our ancestors would have faced tremendous challenges in order to thrive on land and eventually transit across the planet. The more and varied experiences that they undertook and won, the greater the brain and body development experienced with the passage of time. These experiences, coupled with a nutritionally rich diet, would have helped our ancestors grow bigger brains over time.

Many skeptics of evolution often use the development of the eye (much less the development of the more complex brain) in their arguments, arguing that a structure as complex as the eye, which entails the pupil, retina, lens and related structures, could not have come about as a result of evolution. In his book, *"On the Origin of Species"* Charles Darwin notes that the evolution of the eye by natural selection seemed at first glance "absurd in the highest possible degree." However, he went on that despite the difficulty in imagining it, the evolution of the eye was perfectly feasible:

...if numerous gradations from a simple and imperfect eye to one complex and perfect can be shown to exist, each grade being useful to its possessor, as is certainly the case; if further, the eye ever varies and the variations be inherited, as is likewise certainly the case and if such variations should be useful to any animal under changing conditions of life, then the difficulty of believing that a perfect and complex eye could be formed by natural selection, though insuperable by our imagination, should not be considered as subversive of the theory.

In recent years, ample evidence supporting the evolution of the eye from a simple and primitive photoreceptor to a highly evolved, complex structure has come to light. This, despite skepticism by some that there has not been sufficient geological time for the evolution of the eye, much less the brain. A study by a pair of Swedish scientists, Dan Nilsson and Susanne Pelger, suggests that only a small fraction of that time would have been sufficient for the evolution of the human eye. They note that when referring to the eye, one implicitly means the vertebrate eye, but the serviceable image-forming eyes have evolved between 40 and 60 times, independently from scratch, in many different invertebrate groups. Among these 40-plus independent evolutions, at least nine distinct design principles have been discovered, including pinhole eyes, two kinds of camera-lens eyes, curved-reflector ("satellite dish") eyes, and several kinds of compound eyes. Nilsson and Pelger concentrated on camera eyes with lenses, such as the well developed eyes of octopuses and vertebrates.

Nilsson and Pelger note that the first fossils of the eye date back about 540 million years (Cambrian period). A conservative estimate of the time required to evolve the vertebrate eye from a basic receptor cell is 364,000 years. More likely, they estimate that 250,000 years would have be required for the evolution of the vertebrate eye, which is a relatively short period in geological terms. Indeed, if the eye can evolve independently forty or sixty times, the brain can also evolve and lead to the development of the human brain[10].

Returning to our ancestors, due to the challenges faced in living on the ground rather than in trees, it is likely that our ancestors partnered with some animals to help them survive this transition. As they struggled to find food and shelter and to keep wild animals at bay, wild dogs (wolves) would have been among their partners. With the passage of time, these animals likely stayed near our ancestors' camps (for bones and food scraps) and kept other wild animals (e.g., lions, tigers, leopards, cheetahs) from terrorizing and preying on them. In return, our ancestors would have shared food and shelter with them. This symbiotic relationship would have been mutually beneficial to both parties. Wild cats may have also joined the human camps as rats and mice would have followed the human trail of food. With the passage of time, wild dogs and cats would have lived among our ancestors, including by helping them stay warm during the cold seasons by sleeping at their feet and serving as "blankets" to keep them warm. Eventually, our ancestors would have domesticated wild dogs and wild cats to what they look and act like in modern times. Today, perhaps given the long history of humans and such animals living and surviving together, it is no surprise that dogs are considered man's best friend and cats are common household pets (or at least they live in close proximity to humans). It seems that the bond between dogs, cats and humans is a long standing one.[11]

In the process of having animals in their shelters and in

such close proximity, our ancestors probably learned many things about them. At some point, given their expanded brain capacity, humans would have been able to formulate cause and effect relationships and probably came to understand the flow of life. Our ancestors may have learned what gives rise to puppies and kittens by watching dogs and cats copulate and later give birth. Eventually, irrespective of how they made this association, our ancestors would have developed the brain capacity to understand that copulation is necessary for bearing offspring. In the process, women would have understood when they became pregnant and were eventually able to make the association between copulation and pregnancy. Perhaps women were the first to figure out this association (or at least the first to confirm this association) as they are the ones who become pregnant. Eventually, this knowledge would have led humans to cover their genitals, i.e., in order to control the copulation process to produce the desired offspring rather than producing random offspring as in the animal world. Regardless of how they gained this knowledge, at some point, humans developed the brain capacity to understand that copulation leads to offspring and thus started to cover their private parts and began to actively influence their progeny (along with actively breeding of dogs and cats in their shelters and camps). Covering their private parts, and setting up a social structure for producing and rearing offspring, allowed humans to influence and, to a sizable degree, shape future generations of humanity unlike any other animals.

Perhaps this is the genesis of the Adam and Eve story in the Garden of Eden. Eve may have achieved awareness of the relationship between copulation and offspring, at least in her own body when she became pregnant. She would have shared this "knowledge" with her partner and her social group. Realizing this, they would have felt shame in exposing themselves to others in their midst and covered their genitals. With this "knowledge," rather than remain ignorant and operate only on their innate animal nature, eventually our ancestors would have actively decided to bear and raise their offspring together. This level of understanding of the flow of life would have been possible because the human brain had evolved sufficiently to formulate such cause and effect associations and to decipher how children come into being.

Through this knowledge, humans thus began the journey into understanding the workings of the natural world and into truly becoming *Homo sapiens*. This knowledge would have allowed humans to begin carving a new path for their existence. A life where humans make a concerted contribution toward their life and their progeny rather than just existing in the natural world and operating solely in accordance with their innate animal nature. This was humanity thinking for itself and taking control of its life (and its progeny) and thus being "banished" (i.e., carving a separate path) from the Garden of Eden, where nature alone is in control and all other animals exist and operate on the basis of their innate instinctive drivers. Humans had thus taken this key piece of "knowledge" from the tree of knowledge, which would position humanity along a new path. A new path for humanity had been drawn, eventually leading to the humanity that exists today. In this context, it is no surprise that about 170,000 years ago, i.e., shortly following the dawn of humanity, our ancestors were generally wearing clothing.

The knowledge that copulation leads to offspring would have also led humans to eventually live in well defined social groups (and wear clothing to at least cover their genitals) where copulation is controlled to produce the desired offspring (and at the desired age). This discovery would have also promoted greater communication among humans in order to share this knowledge and to incorporate it into their social structure. With the passage of time, the children that the humans bore would be raised together by the two parents, for an extended period of time. Eventually, this would lead to the establishment of the institutions of marriage and family. Thus establishing a social construct within each social group regarding how, with whom, and when an individual would "marry" and copulate to produce and rear offspring.

This approach to influencing and controlling their offspring would have ultimately allowed early humans to evolve along a different path than other animals. Traits valued by specific social groups would have been allowed to flourish (through pair bonding and marriage) and result in progeny carrying these traits. Men and women would have been able to have some influence on the traits (both physical and mental) that they would want in a partner. Unlike the animal world, where mainly physically dominant males reproduce, human males with valued mental traits (e.g., intellect and empathy) but seemingly non-dominant physical traits would have also been allowed to reproduce. Similarly, women who were not necessarily physically dominant females, but possessed the intellect and empathy needed to rear good offspring and maintain a strong family structure, would have been selected for alliances. Ultimately, humans possessing a balance between the desired physical and mental traits would have be preferred and prevailed. This new approach to progeny allowed humans to "self breed" and to bring "knowledge" and "caring" into the mix of valued traits.

With the passage of time, humans would "evolve" such that their animal nature would be increasingly balanced with their learned nature. Humanity would no longer operate just on its innate instincts, but rather on its innate instincts balanced with intelligence and emotions such as empathy. This is not to say that animal instincts, which are innate and operate on the senses (sight, sound, taste, touch, and in particular, smell driven by pheromones) are bad or devoid of caring. Rather, the point being that humans were no longer at the mercy of just their innate animal instincts in existing in the world. Intellect and empathy were intimately weaved into the fabric of human traits. Once these aspects of humanity had taken root, language and communication became even more crucial within their social groups, e.g., for discussing ideas, sharing knowledge and expressing compassion and even "love" toward others. Thus, with increased focus on language and communication skills, the human brain would have further evolved over time. Thus beginning the journey to becoming *Homo sapiens* or wise men.

Humanity's desire to use its brain and its perspectives on the natural world to shape its existence, to understand life and the universe, to fight despair, to make life comfortable, to express compassion and true love, and to reduce hardship and suffering, continues to this day. Humanity had thus carved a new role for its existence in the natural world. Humans no longer acted on just natural instincts. Instead, humans acted by applying their natural instincts balanced with knowledge and caring.

Looking to today, *Homo sapiens*, by using their brains, have managed to control their environment to such a degree that we are now spread throughout the world and wield incredible influence on the planet. Humans today continue to strive to balance instincts with knowledge and caring within our human cultural/social framework. We are part of the natural world and operate within it, but we use our "brains" in addition to our innate nature in making decisions, including mating decisions. As noted earlier, humans often choose to marry only when they are ready to settle down and are able to care for their young. It is not the same as the animal world where mating is based on innate instincts. In the animal world, often only the physically strong and fit are able to mate and their instincts and circumstances lead them to focus on protecting themselves at any cost. Humans, being a strong and yet thinking animal, have evolved to use instincts as well as knowledge and caring to protect the weak (along with the strong). Humanity continues to be a part of the natural world and to evolve within it, but humans operate as knowledgeable and influential participants in the world rather than just exist in it.

Humanity's journey has ultimately led it on the path to making great discoveries and inventions in all fields of study —e.g., medicine, science, math, and physics. Humanity's knowledge of the natural world and its inner workings is vast, but it is still in its infancy. There is still a lot to learn and to find ways to use knowledge to benefit our existence and environment (rather than to its detriment). It is often that, with each major step forward that humans take, there are one or more (hopefully smaller) steps backwards. Regardless, over the thousands of years, humans have learned a lot about the natural world within which they live and have found ways to make their life more comfortable and meaningful.

6 BALANCING INSTINCT WITH INTELLECT

We have danced the ultimate dance with evolution to create ourselves.
— *N. K. Sachdev*

Humans have actively shaped their world and their life.

Humans have also helped shape their brain and body in the process. This has been achieved in the process of integrating knowledge and compassion in concert with evolution and the environment. This is not to say that other creatures do not shape their bodies and brains—they clearly do. However, humanity has played a truly impactful role in its own development. Humanity has danced the ultimate dance with evolution to become what humans are today—*Homo sapiens* or wise man.

What does this mean? Let us begin by examining nature.

Nature seeks to achieve balance and continuity, including in the varied forms of life it can sustain. For example, in nature there is balance in terms of how many animals of a various kind exist in a given environment. The number of mice in any one area depends critically on their food sources and their reproductive rate, offset by the number of predators (e.g., cats, snakes). Nature, in a particular area, can only sustain a certain number of mice given the critical variables for their survival and reproduction. Thus, there is balance between the number of mice and cats and snakes, which nature seeks to preserve.

Proceeding along this thought process, mice do not control their offspring to the degree humans do. They operate on instinct or in a "programmed" approach and thus during certain times of the year, they copulate and produce offspring. This is not to say that there is no selection process when mice mate. Indeed there is selection of mates and mating rituals. However, their innate drivers serve as controlling factor in such situations. Thus, their progeny, like them, are driven by innate instinctive drivers. Humans on the other hand, having developed the intellectual capacity to understand mating and copulation and its impact on progeny, have adopted various methods to actively influence their progeny. In the process, humanity has added knowledge and compassion as key elements, in addition to innate instincts, into the mix of traits for their progeny.

This understanding of the animal world may prompt some to ask if animals only operate on their innate drivers; that is, do they think and have feelings? Indeed, animals have brains and think—they make strategic decisions and operate based on the capacity of their brains. Feeling is with the senses and as animals are tethered to their senses, they feel based on the acuity of their innate senses.

Along the same line of thought, based upon instincts, animals are "programmed" to focus on their basic needs for survival. For example, if they find food, they must feed themselves and their progeny first and store some for the future before sharing any food with other members of their group. This is not surprising as their primary goal is survival and reproduction to keep themselves and their progeny alive.

Humans have the same basic goals, but we have used knowledge to develop an array of strategies and tools for meeting these needs. Thus, given the capacity to meet these needs with comparative ease, humans are able to, and have the capacity to, focus on higher level needs, wants, and goals. For example, most humans have access to sufficient food to feed themselves and their entire family. Food storage is also not an issue for the majority of humans. Thus, humans have the luxury of broadening their goals to include, e.g., seeking knowledge and understanding by reading/writing, attending school, etc. Many humans are also in a position to help others who may need food. Moreover, humans are able to make sacrifices and charitable contributions to care for others in their family, their community, their country, and the world. Other animals also make sacrifices, but humans have the capacity (and the resources) to far exceed any other animal in this regard. Bottom line, humans have found ways to be more than just creatures eking out a daily existence. Humans have, over time, found ways to increase the use of knowledge, in particular knowledge of the natural world and how nature operates, to make their lives more comfortable, to shape their existence, and to improve the health and well being of humanity.

What humanity has achieved in the last 200,000 years is truly astounding. This has been accomplished by humans, by building and applying knowledge and compassion in conjunction with their natural instincts to enhance the standing of humanity, and to try to make sense of the world in which they live.

7 CARVING A PATH

"Man will become better when you show him what he is like"
—*Anton Chekhov*

Humans have played a pivotal role in carving their path in the natural world. In the process, humans have made great discoveries about our world, our planet, our galaxy, and our universe and beyond. We humans are the ultimate toolmakers and we have reached far beyond beyond the physical tools. We have used "intellectual tools" from various fields of study, e.g., math, physics and astronomy, to decipher the laws of nature and to understand how nature and the universe operate. We have advanced our understanding on many spheres—for example, in the fields of medicine, science, art, music, computers, etc.

Humans have also established economic and political systems to help humanity thrive. In conjunction, humans have established institutions and governance practices, policies and laws, as well as a host of supporting rules and procedures to guide humanity. Concurrently, humans have added meaning and beauty into their lives in many ways through art, music, dance, architecture, etc. These are just a few examples of the vast array of human accomplishments. Humanity has made the great leap from a harsh daily existence in the wild to a more beautiful existence for itself.

In the process of carving a path in this world, humans have established ways to support humanity's existence in this world. Imagination, creativity and innovation have allowed humans the ability to expand into environments that our relatively frail bodies could not have otherwise allowed. As the number of humans have grown and expanded across the globe, varied approaches for applying knowledge have been instituted. Humans have also had to fight despair and to deal with the challenges that accompany a complex brain-based existence. On balance, humans have reaped the rewards for their efforts in terms of a relatively comfortable existence (as compared to other animals). Unlike other animals who continue to operate on their innate nature and must deal with daily survival, most humans have transcended this basic level of existence and are more aware of the world, of their surroundings, and of what is happening. Humans are also able to contemplate what the future may hold and understand, often in intricate detail, many aspects of the past (both near term and the more distant past) to a degree unmatched by other animals.

Humans have also developed the capacity to truly love at a level beyond other animals. This, despite the ability of animals to express affection and love in truly pure and innocent manner. For example, modern humans have the capacity: to empathize with those facing unfortunate circumstances; to feed the hungry, to care for the decrepit and infirm, to love the world (despite all the hate that exists within humanity); and, to find a way to forgive (eventually) even their worst enemies. Whereas, in the animal world, the natural instinct is to look after oneself and one's progeny, and then one's group, and to destroy all enemies immediately. Humans have evolved to be the thinking species; thus, we balance our natural instincts with intellect and empathy. Mahatma Gandhi expressed the human capacity for love quite eloquently when he stated that "real love is to love them that hate you, to love your neighbor even though you distrust him."

Humans also have the capacity to learn and to teach their fellow humans at a level that vastly surpasses all other animals. According to Kevin Laland, a behavioral and evolutionary biologist, human accomplishments "derive from an ability to acquire knowledge and skills from other people." Many other species innovate, e.g., chimps open nuts with hammers, and crows have been seen to place nuts on roads so that cars can drive over them to crack the nuts for them. But, Leland notes, humans are unique in their capacity to teach skills to others through the generations with enough precision, e.g., to study distant stars, to build skyscrapers, or to go to the moon. Indeed many animals teach survival and food acquisition skills to their young. However, humans learn and teach at a much more advanced level. In analyzing knowledge and its transmission across humans, teaching and learning are key to human advancement. Copying is also more advanced in humans. Many animals mimic others, but humans have developed the capacity to not just copy but to copy well.

Human communication mechanisms (e.g., language) are

much more advanced as well, which helps to copy well. Teaching and learning benefitted our ancestors greatly in terms of survival and food procurement. Leland further notes that it is the human culture that has given humanity its brains, intelligence and language rather than the other way around. He defines culture as behavior patterns shared by members of a community that rely on socially transmitted information.[12] All animals form groups, build bonds and work toward mutual success. What sets humans apart is their complex culture. The human culture has spawned an existence wherein the level and complexity of human thought, learning, communication, and replication of information and ideas to address problems and meet needs, wants and interests sets us apart from other animals.

Given our complex culture coupled with our thinking brain, we have moved beyond the meager daily survival, hand-to-mouth existence and found ways to thrive rather than just survive, and to love deeply from the inside rather than just the outside. We are kinder and less sexually coercive than most of our nearest relatives. For example, human males help rear their young for an extended period of time unlike other apes and unlike even the bonobos who appear to be more peaceful than humans. On the other hand, humans also possess an intense capacity to hate and cause harm and destruction to a greater degree than any other animal. Humans have the capacity to do incredible justice and, on the other hand, to wreak vast havoc. The suffering caused by the ravages of war serves as a clear example of the harm and havoc attributed to humanity. The path we must chose is intuitively obvious. It is the path paved with knowledge and love.

8 Humanity's Spirit

You are not here merely to make a living. You are here in order to enable the world to live more amply, with greater vision, with a finer spirit of hope and achievement. You are here to enrich the world, and you impoverish yourself if you forget the errand.
—*Woodrow Wilson*

The number and array of human accomplishments is vast. It is amazing what humanity has learned and created in its short existence in the history of this planet. In the process, humans have also had to learn some very hard lessons. With each new system or innovation that humans undertake, often challenges emerge along the way. There are often hard lessons to be learned and often there are adverse consequences that need to be addressed. In the process, humanity has been humbled by the power of nature and the universe. Indeed, humanity has a long way to go to more fully understand the interconnectedness that exists within nature as well as the delicate balance that nature strives to achieve. In the end, nature's goal is balance and continuity, including that of life in all its richness and diversity.

It seems that humans seek to control nature, but this approach has placed nature and its inhabitants in a precarious situation. According to Satish Kumar, noted Indian philosopher and activist, "we have to shift our attitude of ownership of nature to relationship with nature. The moment you change from ownership to relationship, you create a sense of the sacred." According to Kumar, humans need to build a partnership with nature focused on reciprocity and interdependence.

Humanity's primary aim has been to thrive and to use nature's resources to advance human existence. This approach has challenged nature and the very environment within which humans and all living things on earth must subsist. It is now projected that human activity is causing unprecedented mass extinctions. Humanity's challenge is to place priority on helping nature maintain its intricate balance and to respect nature's need for diversity and variability of life forms for its survival.

Imagination, curiosity and ingenuity have been among the key hallmarks of humanity's success. Humans have developed and applied knowledge and understanding of many aspects of nature and its resources and found ways for humanity to thrive. Humans have the capacity to innovate and outthink all other animals. Humans have also applied their knowledge to understand life, the world, the universe and beyond. Humans have deciphered some of the key underlying forces and laws of nature and applied them to their benefit. As a result, humanity has formulated complex and elegant theories, principles, and approximations such as the theory of relativity, quantum mechanics, the Big Bang, the age of our universe, star formation, the Laniakea Supercluster to which our galaxy is connected along with 100,000 other galaxies, and future trajectories of planets, galaxies, etc. We have also deciphered the genetic code of life, mapped all known animals within a phylogenetic tree, and traced all known plant life. With the advent of artificial intelligence, humans are on the path to expand knowledge at astonishing speeds. Humans are the only animal today with the capacity to seek, formulate and comprehend the depth and breadth of such knowledge.

At the same time, humanity has placed its mark in the world in more basic and yet important ways. In the process, humanity has given greater meaning to life. For example, looking at our past, at some point the human population started to name each newborn "life" with a given and a family name, and to maintain official records, e.g., birth, marriage, divorce and death records. Humans also started to record historical information on a timeline. Maintaining historical records of varied sorts and managing vast quantities of information and knowledge bases, in particular about human activities, is a true accomplishment. No other animal has the capacity to achieve anything similar for members of their species. A vast array of tools, ideas, and information attributed to humanity exist today, and the list continues to grow exponentially. In the process, unlike all other living things, humanity has given greater meaning to life and, in particular to human life.

We have also established universal rights for all humans (which unfortunately require much work to be achieved). Humans have given some meaning to all life (e.g., by gaining knowledge of, utilizing, and setting up a nomenclature for all living things, including plants, animals, bacteria and fungi) and naming and mapping all things within our purview, e.g., roads, lakes, mountains, the earth, the other planets, the sun and the moon, the stars, etc. There are many, many other ways in which humans have made life more meaningful and richer. In short, humanity has risen above all life forms and enriched, in particular, the welfare of humanity.

Despite the overwhelming accomplishments, humanity has much work to do. Many members of the human race continue to live with insurmountable challenges and lack basic human rights. As a result, their lives hang in the balance. Many others are not valued for what they can offer the world and are held back while their well-to-do compatriots continue to prosper financially and socially, often generation after generation. Aspects of nature and the environment are also severely challenged as a result of human activity. The future of many animals hangs in the balance, in large part, due to human activity.

As our ideas and knowledge base grow and become more interconnected through technology, our belief systems, in particular in relation to religion, need to coalesce to be more united and encompassing. Our universe is larger than we could have ever imagined and there are possibly multiple universes in existence. Yet, many humans have not applied this knowledge to accept that all of humanity is a very small part of a much larger existence; that our collective "deity" spans a vast existence; and, that it is the divinity within us humans that needs to be more fully awakened and expressed. It is humans who have enhanced the welfare of humanity. It will take our collective efforts working together to further the welfare of humanity and to focus on the welfare of our planet.

The future of our planet needs careful consideration. We need to build the foundations of an existence where humans and other living things can coexist more peacefully. The fear of war and chemical and biological weapons threaten both human and non-human life. Knowledge is a double-edged sword and its vector needs to be directed toward supporting nature and life. This can be best accomplished by humanity, by understanding our past and being cognizant of the pivotal role our ancestors have played in shaping our existence, and, by using our human intellect and empathy to avoid a major catastrophe.

In addition, there are other threats that could compromise the future of humanity and life on Earth. These include the possibility of an asteroid from outer space hitting large swathes of the planet. There is also the concern about solar flares and other harmful agents, e.g., chemicals, gases or other components, seeping into the earth's atmosphere and compromising life forms. A host of other concerns could also make life on earth challenging or perhaps inhospitable. This is an area where humanity could play a pivotal role in preserving life on the planet and beyond. Advancements in scientific research could help humans play this critical role.

The human capacity to meet challenges, often seemingly insurmountable, is a reflection of man's indelible spirit. Humans have worked hard to bring humanity forward. Humans have stepped away from a meager animal existence and used their brains to carve a more meaningful, comfortable and beautiful existence. It is an existence where humans: seek to understand the interconnected nature of reality; give greater meaning to their life and to all life in general; fight for justice for all; and, hope to expand love over brute hatred. It is this noble spirit of humanity that needs to be expanded to stem the conflicts and the cruelty—both among humans and between humans and the environment —to build the foundations of the future.

9 Human Potential

A man is but a product of his thoughts. What he thinks he becomes.
—Mahatma Gandhi

Humanity's future rests on its self-realization and expansion of consciousness. In the process, humanity will better understand its interconnectedness with life, nature, the planet and beyond. Understanding its interlinks with other life forms, humanity must become a true shepherd of the planet and its resources to help sustain it. This includes helping sustain humanity itself and more fully utilizing the skills that have been bestowed upon its members.

Humanity has danced the ultimate dance with evolution to create itself and to shape its world. We are nature's novel experiment with the thinking animal that helps shape and mold its own existence. The bounty that has been created by and for humans is prodigious. Humans have exhibited remarkable ability to innovate in response to their needs and desires. In the end, humans have also utilized their skills and abilities to make their existence sweater and more comfortable as compared with other animals. Humanity has thus given greater meaning to human life and to life itself.

Humans have advanced capabilities compared to other animals. However, humans are not distinct and separate from other animals. Various animals exhibit many of the same capabilities as humans, although some of their capabilities are expressed at, and operate at, a more basic level. This itself reflects man's evolutionary linkages with other animals. For example, elephants have brains similar to humans, including in the number of neurons in the brain cortex, and general brain connectivity and structures. If only they could walk upright and had hands. Defending the uniqueness of human cognitive traits is also challenging as many animals, from birds to chimpanzees, exhibit comparable cognitive skills. Based on research studies, Thomas Suddendorf notes that there are two distinct features that underlie human capacities—complex scenario building and exchanging thoughts with others.

Where humanity has excelled is in its exceptional ability to face challenges through forethought and innovation. Thus, humanity has the potential to help save the planet. It is in meeting this challenge that humanity can expand its consciousness and achieve greater enlightenment.

The path to such enlightenment calls for humanity to more fully embrace and accept: its evolutionary connections with other animals; and, more fully understand the exceptional role that humans have played in shaping their own existence. In addition, humanity needs to realize the critical need to ensure animal and plant diversity. Humanity has danced the ultimate dance with evolution to help root its own human family tree in the evolutionary garden of life and it must preserve the garden and the special tree of humanity through time. Humanity, rather than controlling nature, must more fully understand the interconnectedness of all things. Humanity must also become more conscious of its interdependence on nature—in practical terms—and truly become the wise human to ensure that intelligent life and the biosphere continue.

Epilogue

The world is full of magical things patiently waiting for our wits to grow stronger.
—*Bertrand Russell*

Humanity has come a long way from its ancestral beginnings. The key to humanity's success has been the ability to transform ourselves into *Homo sapiens* through sheer determination, to apply reasoning to intuit solutions to address challenges and problems, and the ability to inform, educate, influence and expand our social group in the process. These traits, aided by evolution and the environment, have allowed modern humanity to emerge as the dominant force on the planet.

As Thomas Suddendorf has poignantly noted, what makes us human is that we can think about alternative future realities and make choices to affect the future outcomes.

Humanity must build the foundations of a new way of living that benefits humans as well as nature. The challenge is to use knowledge to build the foundations of a stronger future for nature and life. Humanity is also in the unique position to help the planet protect itself, e.g., from asteroids, solar flares and other harmful forces. Perhaps, it in this way that humanity will redeem itself and concurrently ensure its own survival.

ABOUT THE AUTHOR

N. K. Sachdev is a philosopher, a scholar, a student of divinity and a scientist. She enjoys analyzing and reflecting on our world to better understand nature and humanity. She is a vision-oriented thinker with the ability to apply diverse knowledge, skills and experiences. She looks forward to reading, writing often, and further developing her understanding of our world.

Notes

1. Source: *Wikipedia*, Evolution of the Brain.

2. "What does it mean to be human?" Smithsonian, National Museum of Natural History. Humanorigins.si.edu/human-characteristics/brains.

3. Mitchell, Kevin, " Ancient Origins of the Cerebral Cortex." September 2010. Wiringthebrain.com.

4. Choi, Charles Q, *LiveScience*, "Human Evolution : The Origin of Tool Use," November 11, 2009.

5. Geneticists note that a reference to "Mitochondrial Eve" do not imply "the first woman" or "the only woman living at the time," nor "the member of a new species."

6. Source: *Wikipedia*, Y-Chromosomal Adam and Mitochondrial Eve.

7. Source: *Wikipedia*, Mitochondrial Eve (Not a fixed individual over time).

8. Duchene, L. "Probing Question: Why Are Flowers Beautiful?" January 24, 2008. Phys.Org.

9. The amylase enzyme is present in the saliva of humans and some mammals. Amylase helps hydrolyse starch into sugars,

resulting in a sweeter taste as starch in chewed. Research studies of human amylase (AMY1) indicates that the pattern of variation in copy number of the human AMY1 gene is consistent with a history of diet-related selection pressures, demonstrating the importance of starchy foods in human evolution.
https://www.ncbi.nlm.nih.gov/pmc/articles/PMC2377015/.

10. Nilssen, D. and Pelger, S. "A Pessimistic Estimate for the Time Required For An Eye to Evolve," *Proceedings: Biological Sciences*, Vol. 256, No. 1345 (Apr. 22, 1994), pp. 53-58. The Royal Society.

11. The human bond with the horse developed much later in time.

12. Laland, Kevin, "How We Became a Different Kind of Animal," *Scientific American*, pp. 34-39.

Bibliography

Choi, C. Q, 2009. "Human Evolution: The Origin of Tool Use, " *Livescience.*

Dawkins, R. 2004. *The Ancestor's Tale.* New York: Houghton Mifflin.

Darwin, C. 1859. *On the Origin of Species.* London: John Murray.
_____. 1871. *The Descent of Man, and Selection in Relation to Sex.* London: John Murray.

Duchene, L. January 24, 2008. "Probing Question: Why Are Flowers Beautiful?" Phys.Org.

Hirst, K. Kris, 2018. "The Discovery of Fire," *ThoughtCo*: 169517.

Kramer, P. "Walking with Cavemen- 419/1" *BBC*
    http://www.bbc.co.uk/sn/prehistoric_life/tv_radio/wwcavemen/.

Kumar, S., Editor of Resurgence and Ecologists, Resurgence.org.

Laland, K. Date. "How We Became a Different Kind of Animal," *Scientific American.*

Mitchell, Kevin. "Ancient Origins of the Cerebral Cortex." September 2010. Wiringthebrain.com

Nilssen, D and Pelger, S, Proceedings: Biological Sciences, Vol. 256, No. 1345 (Apr. 22, 1994), pp. 53-58. **The Royal Society.**

Prum, R. 2017, *The Evolution of Beauty*. Doubleday: New York, London, Toronto, Sydney, Auckland.

Redford, T, "The Untold Story of Evolution"(April 25, 2011), The Guardian/Science.

Suddendorf, T. 2013, *The Gap: The Science of What Separates Us from Animals*. Basic Books: New York.

What does it mean to be human?" Smithsonian, National Museum of Natural History. humanorigins.si.edu/human-characteristics/brains

Wikipedia (Evolution of the Brain)
        (Y-Chromosomal Adam and Mitochondrial Eve)
        (Mitochondrial Eve "Not a fixed individual over time')

The Spiritual Man, The Buddha

—

www.ingramcontent.com/pod-product-compliance
Lightning Source LLC
Chambersburg PA
CBHW030451220526
45464CB00006B/2480